GRAIN ELEVATORS

GRAIN ELEVATORS

LISA MAHAR-KEPLINGER

PRINCETON ARCHITECTURAL PRESS

This catalogue has been made possible by the generous support of the National Endowment for the Arts and the New York State Council on the Arts. The study was sponsored by the Preservation Coalition of Erie County.

Published by
Princeton Architectural Press
37 East 7th Street
New York, New York 10003

Printed and bound in the
United States of America

Editor
Stefanie Lew

Library of Congress Cataloguing-in-Publication Data
Mahar-Keplinger, Lisa, 1965–
    Grain elevators / Lisa Mahar-Keplinger.
        p.   cm.
    Includes bibliographical references.
    ISBN 1-878271-35-0
    1. Grain elevators—Design and construction.   I. Title.
TH4461.M34   1993
725'.36—dc20   92–29683
                                    CIP

To Eric
For his unerring critical insight,
support, guidance, and love

Table of Contents

Timeless Cathedrals

To those who travel the great highways of the Midwest, silos appear like cathedrals, and in fact they are the cathedrals of our times. Their materials impose the rhythms of this book—wood, brick, tile, steel, concrete—and they mark the passage of time, the slow evolution of a collective work.

The Great Plains of America are vast, and secret are its villages turned inward on their religious sects and antique languages, as if time had stood still. These people were not seeking America, but were escaping from Europe, and in these first wooden silos there is a memory of, and an obsession with, architecture from different parts of central Europe. Over time the silos rose with ever greater assurance and created the landscape of the New World. In abandoning the problem of form, they rediscovered architecture.

From out of the gray landscape of publications on architecture emerges this book that searches for something authentic and definitive. Lisa Mahar-Keplinger analyzes these architectures typologically and in terms of their constructional systems, disguising the secret of their allure. Her photographs, taken as she followed the route of the pioneers, are like black and white etchings. We are struck by the purity of the geometries, the clarity of the construction, the relationship with the landscape. And yet the author's photographs also speak to us at another level: the sky, the shadows, the composition simultaneously reveal and conceal the beauty that we seek.

Somehow the book makes me think of James Joyce, and not just because the author is of Irish descent, but because of the way the landscape appears and disappears within the story. In photographs of the elevator in the landscape [see pages 28–29], we find the fresco of the American countryside constructed of a few essential items: the grain elevator, a few trees and telephone poles give us a scene much like the profiles of the hills in the films of John Ford.

Lisa Mahar-Keplinger has located something that perhaps even she did not expect to find: architecture. In these times of so much mediocrity I rediscover a faith that at times I feel I have lost. This small book teaches us that despite everything, even our profession can participate in the search for truth.

Aldo Rossi
New York, November 1992

Translated by Diane Ghirardo

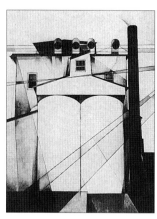

1915

1923

1926

1927

Grain Elevator
Erich Mendelsohn

Towards a New Architecture
Le Corbusier

Minneapolis
Louis Lozowick

My Egypt
Charles Demuth

The Grain Elevator Observed

As a building type, both rural and urban, the grain elevator has provided a source of inspiration for architects and artists alike. From the European architects who first noticed them at the turn of the century to the generations of American artists who still document them, all have continued to renew the meaning of the elevator through their work. American artists, in particular, developed a certain way in which they recorded the elevator. They emphasized the elevator's context, whether rural farmland or urban industrial landscape, and its identity as an American object, thereby setting themselves apart from the Europeans who were interested primarily in the building's form—a quality that transcends cultural boundaries. Both groups of artists, however, strove to extract the spiritual essence of the grain elevator, without altering its integrity as an object, and transformed it from a common vernacular structure to a building of iconic stature. Their work shows us that the grain elevator, although perceived as an industrial object, is intimately connected with the land and the cycles of nature.

The grain elevator was one of the first American building types to receive international attention and influence an entire generation of European architects. Le Corbusier, Erich Mendelsohn, and Walter Gropius discovered the elevator in American trade journals and were drawn to its formal simplicity and honest use of materials. For them, the clarity of the relationship between form and function superseded the elevator's cultural identity, thus they had no difficulty adopting it, both formally and symbolically, as a model for their international vision of architecture. Mendelsohn, in a visit to Buffalo, New York, remarked on the colossal scale of the concrete elevators he saw there: "stupendous verticals of fifty to a hundred cylinders, and all this in the sharp evening light, everything else now seemed to have been shaped interim to my silo dreams. Everything else was merely a beginning."

In response to this European interest in industrial buildings in the United States, American artists began incorporating the grain elevator into their work. During the 1920s these artists, who had been working within the one-hundred-year-old American realist tradition, shifted their approach and began experimenting with abstraction. Influenced by cubism, yet drawing on their realist heritage, the Precisionists attempted to break down the geometry of the subjects they were painting. Rather than strive for total abstraction, however, they focused on emphasizing basic shapes and forms. Their preference for painting buildings and cityscapes distinguished them from their European counterparts who continued to paint the traditional subject matter of still lifes and figures. Unlike Le Corbusier who saw the grain elevator as a universal building form, the Precisionists, like Charles Demuth, treated it as a particularly American type. Demuth's painting of 1924 entitled *My Egypt* sought to both monumentalize and nationalize the elevator. Composed to resemble the ancient pyramids of Giza, the concrete ele-

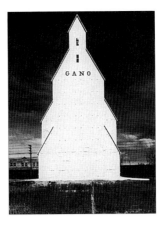

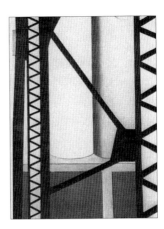

1931

1938

1940

1942

Classic Landscape
Charles Sheeler

Grain Elevator, Everett, Texas
Dorothea Lange

Gano Grain Elevator, Western Kansas
Wright Morris

Grain Elevators from the Bridge
Ralston Crawford

vator is situated in a common American streetscape where telephone pole wires intersect diagonally overhead. Demuth connects the grain elevator to the history of architectural masterpieces while also celebrating its uniqueness as an American object.

Industrialization continued across the rural Midwest during the early 1930s, intensifying the tension between nature and industry. Artists responded with a new sensitivity to the landscape, documenting the grain elevator in relationship to its immediate context rather than as an isolated object. Charles Sheeler, in his painting *Classic Landscape*, harmoniously reconciled industry and nature, grain elevator and landscape. In this new landscape, beauty and work were inseparable. Leo Marx, in *The Machine in the Garden*, describes Sheeler's painting as the pastoralization of the industrial landscape—the refusal to relinquish the pastoral ideal even in the face of industrial reality. "By superimposing order, peace, and harmony upon our modern chaos," writes Marx, "Sheeler represents the anomalous blend of illusion and reality in the American consciousness." As the grain elevator and other industrial structures became more common on the plains, they redefined the traditional pastoral image of the American landscape.

The mid-1930s were troubled by the Great Depression. With the economic pressures, the stabilization of midwestern towns, and railroad growth, the construction of grain elevators tapered. Artists sought to record these effects, choosing the documentary method and the

medium of photography. As part of the New Deal, the Farm Security Administration, a federal agency, sponsored a number of talented photographers—including Dorothea Lange and Walker Evans—to travel around the country and take pictures of farmers and their families. Roy Stryker, the director of the photography unit, defined the platform for the photographers: "a good documentary should tell not only what a place or a thing or a person looks like, but it must also tell the audience what it would feel like to be an actual witness to the scene." The grain elevator was recorded in this way, as a document of current social issues and of human struggle.

What was important about the elevator to this generation of artists was that it revealed something about the lives of the people who used it. Each image was intended to be seen as part of a larger group or series. They were combined to tell a story—either about a particular place and time or about a group of people. Often, a description accompanied the image, drawing out the emotional content of the photograph, as well as identifying its location and date. The FSA photographers succeeded in defining a regional American aesthetic through the understanding of the vernacular landscape. This project was the most substantial effort of its kind ever undertaken by the American government.

As the country came out of the depression during the 1940s, many artists moved away from social concerns to formal ones. Wright

1945

Filling a silo, Kanona, New York
Charlotte Brooks

1948

Rice Silos, Sacramento Valley
Ansel Adams

1949

Untitled
Ralston Crawford

1973

Abandoned grain elevator, Kansas
Frank Gohlke

Morris's photographs illustrate this shift. A writer as well as a photographer, Morris continued the documentary techniques of the thirties, but described his particular approach to photography as one in which he was "drawn to forms that were traditional and impersonal." In fact, almost all of his photographs exclude people (the reason he was not hired as a photographer for the FSA). Morris photographed many structures that were either forgotten or abandoned and in this sense, he ushered in an era of nostalgia for a rural way of life that was quickly disappearing as the cities and suburbs expanded.

When Morris wrote about his experience of the Gano elevator, he emphasized both the elevator's form and its iconic power:

On the crests of the rise, as I drove south, I caught glimpses of an arrow that pointed at the sky, like a rocket on its pad, the moon its destination. As I moved closer I saw the staggered tiers of a grain elevator approximately in scale with the landscape. . . . An almost high noon light, filtered through an overcast [sky], revealed the ripple in the sheet metal attached to the structure's surface. . . . Near the top, appropriately enigmatic, the four letter word G A N O.

Morris's focus is on the building and its poetic implications, but unlike the documentarians of the 1930s, he regards the elevator strictly as an object, not as a representation of social issues.

The 1940s also saw a renewed interest in the experimentations of the 1920s. Ralston Crawford's interest in abstraction continued the Precisionist tradition which he had helped define. His 1942 painting *Grain Elevators from the Bridge* is abstracted to the degree that only two things reveal its subject: the title and the partially visible cylinder bottom. Crawford's photograph of 1949, like Ansel Adams's of one year prior, emphasizes the strong geometry of the building through the heightened contrast of light and shadow. In both works, the elevator becomes a geometric element, not the subject, of the photographic composition.

In the early 1960s, an interest in typology developed, generated by the success of pop art and its use of repetition and common objects. Artists began producing typological series of images as one image no longer sufficed in explaining the complex relationships they wanted to express in their work. The German photographers Bernd and Hilla Becher began taking pictures of vernacular and industrial buildings in series; all variables external to the building—weather, sky, film development, and scale—were controlled. Since the conditions behind each photograph were made uniform, attention was drawn to the subtle differences and similarities of forms. For the Bechers, the documentary approach took on a new meaning: grain elevators were studied in relationship to each other, rather than to people or places, and presented in a straightforward manner.

At this time, a number of important theoretical writings on typology also appeared, and consequently, studies of vernacular buildings

1975

Kreuz auf einem Silo
Walter Pichler

1985

Figure and elevator
Colin Grey

1987

Grain elevator and cemetery
David Plowden

1992

Grain elevator
Aldo Rossi

became more common as historians searched for connections between culture and architectural form. Aldo Rossi, in *The Architecture of the City*, discusses typology and the importance of studying individual buildings in relationship to both history and the city. Type, writes Rossi, "is to be found in all architectural artifacts. It is also then a cultural element . . . ." His interest in vernacular architecture is based on history and the evolution of form rather than function: type is the transmitter of tradition. "In reality," he continues, "we frequently continue to appreciate elements whose function has been lost over time; the value of these artifacts often resides solely in their form . . . ." For Rossi, then, the grain elevator expresses an inherited way of life—what he calls the collective memory of a place—embedded within a specific building type.

By the 1970s and 1980s, this renewed historical interest of the sixties transformed into a fascination with ruins. Most grain elevators, built in the first quarter of the century, were beginning to show signs of age. In response, photographers like Frank Gohlke documented the irony of a building conceived through function, but now abandoned and without apparent purpose. His photograph of a dilapidated grain elevator in Homewood, Kansas exposes the elevator's awkwardness in ruin, making it obvious that it no longer fits into its context. Walter Pichler, an Austrian artist, saw a certain poeticism in this decay: his drawing for the adaptive re-use of a pair of silos in ruin presents these common

objects as sacred artifacts intimately connected to nature and the daily rituals of life. In contrast, Colin Grey's mystical picture shows the crumbling foundations of a concrete urban elevator, but unlike Pichler offers no alternative for its use. Instead he chooses to expose the precarious relationship between man and his environment.

Formally the grain elevator has changed little over its 150-year history yet, it is interesting to note, the perceptions and representations of the building type are extremely varied. From the functionalist admirations of the turn of the century to a nostalgic fondness at the end, the grain elevator has demonstrated its ability to transcend its utilitarian label and exhibit the timelessness of its architectural features. Artists and architects found allure and beauty in the elevator's geometric purity of construction; they found symbolism in its relation to industry and the rural landscape. Their work reminds us that the honesty and simplicity of these vernacular structures continue to teach and to inspire.

The Rural Elevator

The rural elevator, standing anywhere from seven to eleven stories tall, is the most conspicuous structure on the Great Plains. It towers above the average town building of one or two stories and clearly marks the location of each town from miles away. In more densely populated areas elevators and towns are spaced approximately six miles apart, along railroad lines, providing a regular vertical rhythm on the otherwise flat midwestern horizon.

To a farmer, the most important qualities of an elevator are its structural stability, economy, and ability to resist fire. The common wood elevator met the first two conditions but was prone to grain dust explosions from poorly lubricated machinery and to fire from sparks flying off passing trains. This problem led to brief experimentations with brick, tile, and steel during the first two decades of this century, all of which were finally rejected around 1915 in favor of concrete. The wood elevator continued to be popular because it was economical and easy to build—many details were adapted from barn and log cabin construction. Fire danger was reduced and insurance rates lowered by covering the wood exterior with metal or asbestos siding.

The rural elevator is designed to a standard plan and for this reason most elevators resemble one another. Formal variations are the result of regional differences in local building practices and can be seen in the slight proportional variations and differences in roof configurations. The circle, the square, and the triangle are the basis for every construction, both rural and urban.

The Urban Elevator

The urban elevator is larger and more mechanically sophisticated than the rural elevator and was the type most admired by European architects at the turn of the century. These buildings are monumental—a single bin may be large enough to store the annual produce of a hundred farms. Their function is to receive grain from the smaller rural elevators and store it until it can be sold to processors. Most of these elevators also have the capacity to sort and clean grain. The machine floors house the equipment and are located either on top of the bins or to the side in separate towers. When an elevator is situated directly on the water it has additional towers along the water side of the facade called marine legs. The marine legs support the elevating machinery that lifts the grain from the barges into the building.

Urban elevators are positioned near railroad centers or waterways but are rarely an integral part of their urban landscape. Placed on the edge of the city, they have little effect on its organization unlike their rural counterpart which defines the structure of the town and is the center of its activity. Urban elevators were built by engineers rather than local builders and for this reason they are generally unique in plan. Almost all existing urban elevators are constructed of concrete although there are examples built in all of the standard materials used in rural elevators. The most outstanding collections of historically significant urban elevators are in Minneapolis and Buffalo, where there has been a tradition of talented local engineers. Both cities were important centers for the international distribution of grain.

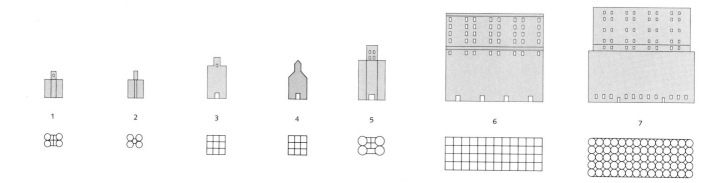

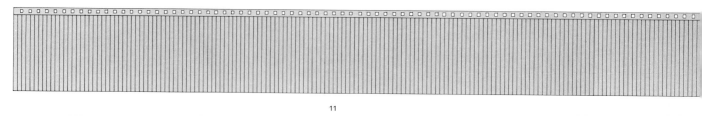

Scale

The primary difference between the rural and urban elevator, besides location, is size. The rural tile, steel, and brick elevators are the smallest with a height between 45 and 60 feet. The rural wood elevator stands at approximately 65 to 75 feet high (in North Dakota and Canada they can be as tall as 110 feet), while the rural concrete elevator stands between 85 and 115 feet high. Early examples of the concrete type were closer in size to the rural wood elevator—the type from which it developed. Over time builders began to understand the particular qualities of the new material and the size of the rural concrete elevator increased. The wood tile, brick, and steel urban elevators, although formally unique, are all quite similar in size, being 300 to 375 feet long and 120 to 130 feet high. The concrete urban elevator, however, can be found in a wide range of sizes, from the 550 foot elevator in Brooklyn, New York (figure 10) to the 2,600 foot elevator in Hutchinson, Kansas (figure 11), which is the longest in the world at the length of nearly ten New York City blocks.

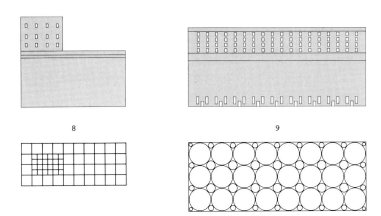

8

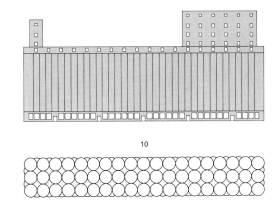

9   10

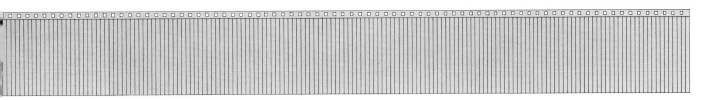

11

0   50   100   200   400   600

WOOD

THE RURAL ELEVATOR

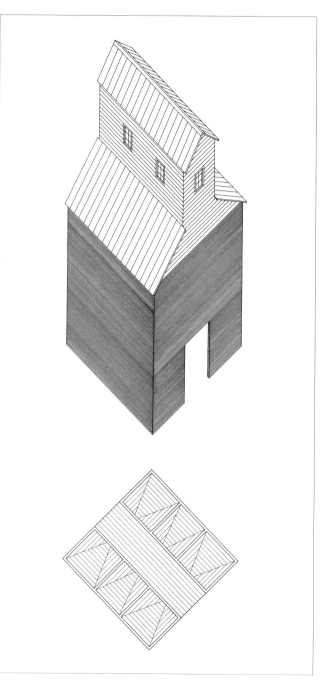

Cribbed Construction

Two different methods are used for the construction of rural wood elevators: cribbed and studded. The cribbed method is the most common technique and the most structurally stable. Walls are constructed with 2x10's, 2x8's, 2x6's, and 2x4's laid flat in a rectangle or square (larger boards on the bottom, decreasing in size near the top), and held together with large metal spikes. The walls interlock like logs in cabin construction. The resulting form is so self-contained that additional storage facilities must be placed in separate annexes alongside the original structure rather than attached directly.

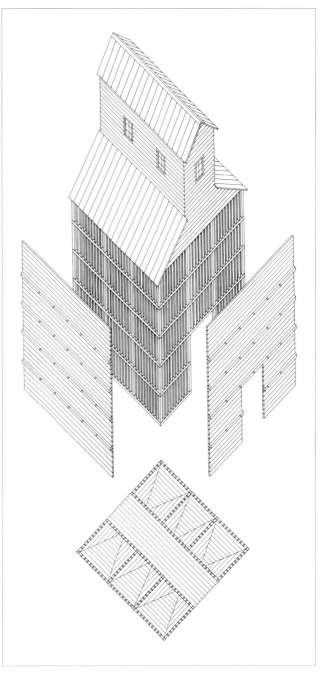

Studded Construction

The studded elevator uses traditional balloon frame construction and is less expensive to build than the cribbed one. The introduction of standardized lumber and the westward expansion of railroad lines made the construction of these buildings possible. The exterior is nearly identical in appearance to that of the cribbed elevator except for the horizontal bands that run around the structure. The bands are made of wood penetrated by tie rods that extend through the elevator interior to support the bins. Almost all rural wood elevators are covered with metal siding or asbestos to protect them from fire.

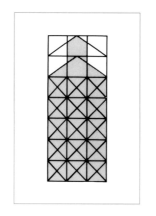
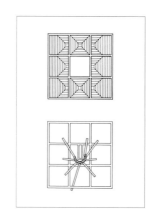
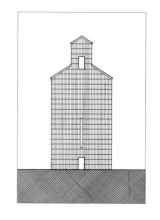

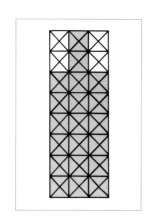
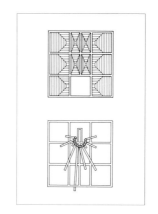
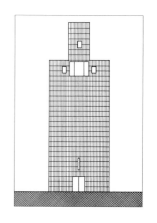

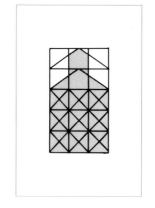
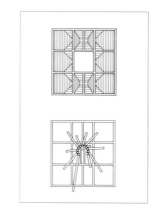
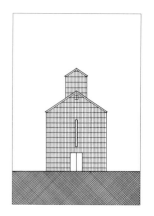

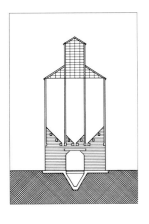
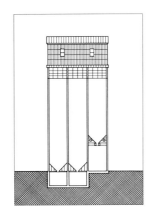
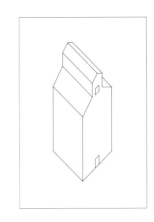

Cribbed Construction

Variations in the cribbed elevator are found mainly in the different roof configurations. The three types presented here are, from top to bottom: the triangular cupola, the rectangular cupola, and the set-back triangular cupola.

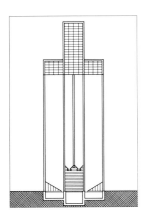
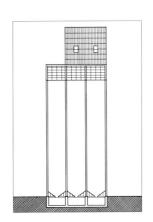
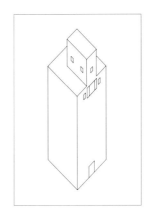

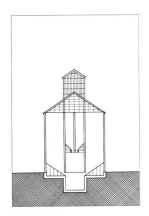
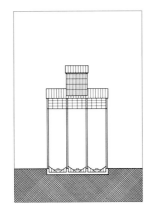
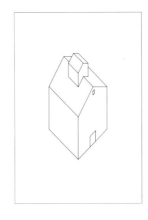

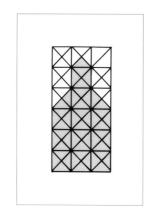
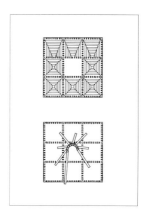
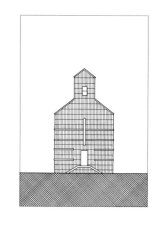

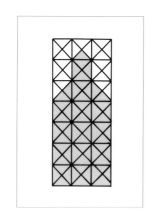
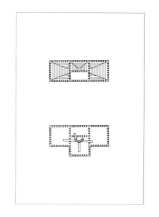
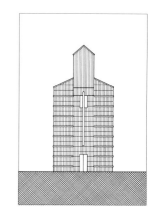

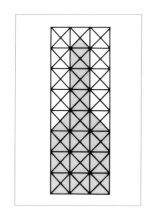
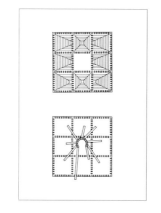
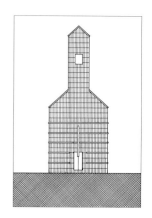

24

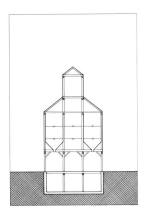
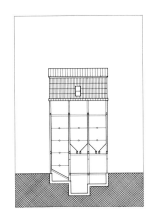
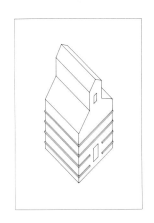

Studded Construction

Variations in the studded elevator are found in the proportional relationship between the cupola and bin structure as well as in the appearance of the horizontal bands that run around the structure. Studded elevators are found primarily on the southern plains in such states as Kansas and Nebraska.

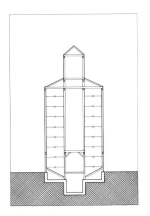
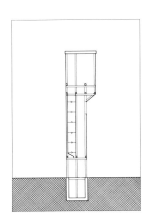
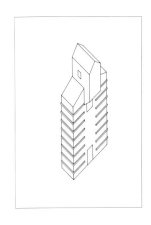

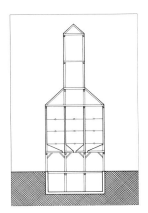
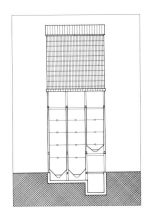
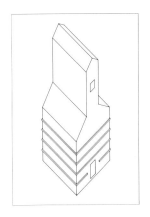

19	20	21	22	23	24
25	26	27	28	29	30
31	32	33	34	35	36

The Elevator in the Landscape

Rural elevators are spaced at regular intervals at a distance that most efficiently serves the local market. In a more populated state like Kansas, they may be as close as every four miles; in North Dakota they are as much as fifteen miles apart. As one elevator comes into view the next can be seen in the distance marking the next town. The elevators shown here are near Kinsley, Kansas. The concrete elevator in the distance is in Offerle.

BRICK, TILE, AND STEEL

THE RURAL ELEVATOR

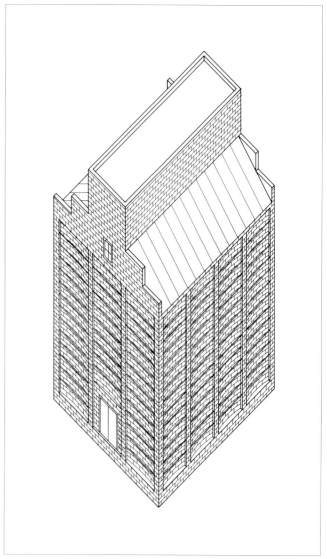

The Rural Brick Elevator

The brick elevator has the most traditional architectural detailing of all the grain elevator types, however it is a type scarcely found today. With its rectangular bins, the elevator in the photograph above shows the typical form of a rural brick elevator, although many have been constructed with circular bins. The elevator in the drawings (above and opposite), located in Rushford, Minnesota, was demolished in the late 1970s. It shows an elegant solution used to counteract the relatively low tensile strength of brick. Curved brick walls placed with the concave side toward the exterior increase stability while tie rods run across the walls and through the columns to increase tensile strength.

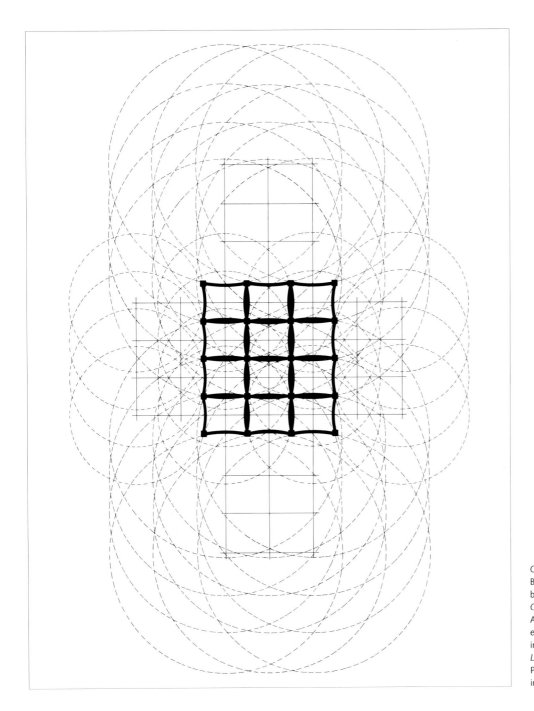

Opposite, left
Brick elevator with rectangular
bins in Andover, South Dakota
Opposite, right
Axonometric of rectangular-binned
elevator with concave exterior walls
in Rushford, Minnesota
Left
Proportional study of brick elevator
in Rushford, Minnesota

The Rural Steel Elevator

The rural steel elevator had two very different periods of development. The early steel elevator, pre-1930, had a simple arrangement of circular bins with either a headhouse mounted on top or a separate steel-frame work-house annexed alongside the bins. This type is extremely rare and, like the tile and brick elevator, was built only for a period of a few years, until the introduction of reinforced concrete. During the 1940s, however, steel bins became popular as additions to rural wood elevators. Farmers were producing an enormous surplus of grain, and in order to assist them, the government began supplying steel bins at low prices. Soon the bins were being used alone and added to until they formed a complex large enough to be classified as an elevator. In this new type, machinery was placed on the exterior of the building or in an adjacent structure rather than inside. This new approach emerged primarily in an effort to reduce the risk of fire and grain explosions, and to facilitate the addition of new bins.

Right
Steel elevator in Kinsley, Kansas
Opposite, clockwise from top left
Tile elevator in Danville, Kansas; tile elevator in Harper, Kansas; tile elevator in Sharon, Kansas

34

The Rural Tile Elevator

The tile elevator, unlike the brick elevator, was always built with circular bins. Tiles were manufactured in a limited number of curved shapes which predetermined bin size and dictated a cylindrical form. The restricted bin size was considered a great inconvenience and often led to wasted space because the bins could not be combined in an efficient manner. Eventually, tile construction was rejected for its structural instability and susceptibility to fire damage. Tiles were also difficult to join, resulting in leaky bins which destroyed the grain. Tile elevators are quite uncommon; most are located in Kansas and Nebraska.

The Search for a New Material

The exorbitant cost of replacing fire-prone wood elevators at an average of every four years, combined with high insurance rates, led to experimentation with alternative materials: brick, tile, and steel. Steel was considered the most promising of the three because of its success in railroad construction. It did not take long, however, to realize that it, too, was an inappropriate material for the storage of grain. Steel, a poor thermal insulator, proved incapable of providing protection for the grain through the harsh summers and winters. None of these three materials were used for more than a few years as builders found applications for concrete in elevator construction.

CONCRETE

THE RURAL ELEVATOR

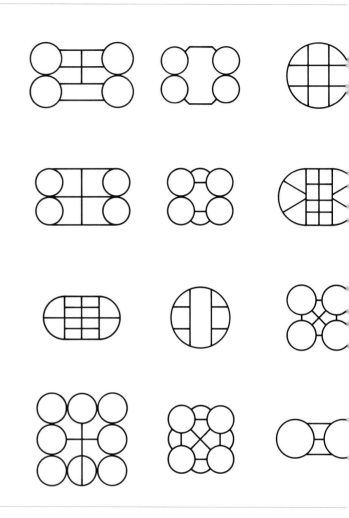

The Rural Concrete Elevator

The rural concrete elevator, which stands thirty to forty feet taller than the rural wood elevator, is easily identified from a distance. The majority of concrete elevators have circular bins, though square-binned examples can be found. The square-binned concrete elevator appears to have emerged as an imitation of the plan and general form of the wood elevator and is similar in size. Skilled labor was not readily available in rural areas; consequently, builders relied primarily on their knowledge of wood elevator construction. As builders became familiar with the material, innovative forms began to emerge, as illustrated in the plans above. While many geometrical variations can be found in plan, all attempt to solve the same problem: how to connect one cylinder to another without waste of space or structural instability. Rural concrete elevators are found in most states on the Great Plains, but are less common in the far north where cribbed elevators predominate. Depending on the region, the rural concrete elevator is either painted white or left its natural gray color.

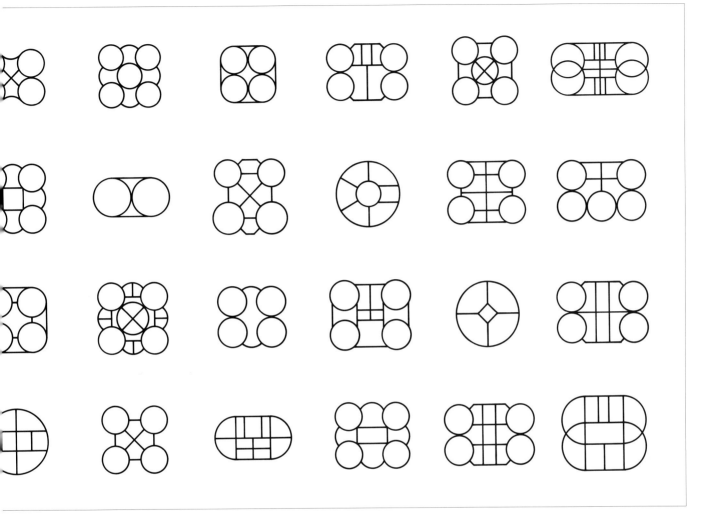

1	Kankakee, Illinois	10	Hite, Washington	19	Vandalia, Missouri	28	Pesotum, Illinois
2	Hagener, Illinois	11	Ottawa, Illinois	20	Cuba, Illinois	29	Cottonwood, Indiana
3	Oneida, Illinois	12	Culbertson, Nebraska	21	Prototype	30	Taylorville, Illinois
4	Hennessey, Oklahoma	13	Texhoma, Oklahoma	22	Springfield, Ohio	31	Olathe, Kansas
5	Omaha, Nebraska	14	Los Angeles, California	23	Washington, Ohio	32	Chicago, Illinois
6	Menasha, Wisconsin	15	Mexico, Missouri	24	Washington, Ohio	33	Malta, Illinois
7	Winfield, Kansas	16	Sheffield, Illinois	25	Oxford, Indiana	34	Evansville, Indiana
8	Mulvane, Kansas	17	WaKeeney, Kansas	26	Hodgenville, Kentucky	35	Pratt, Kansas
9	De Smet, South Dakota	18	Ashkum, Illinois	27	Greensburg, Kansas	36	Talmage, Nebraska

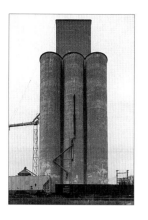
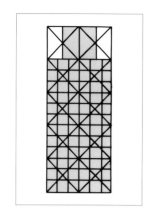
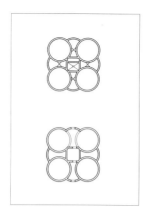
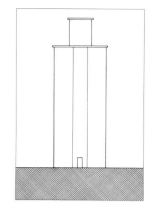

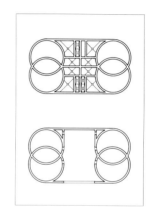
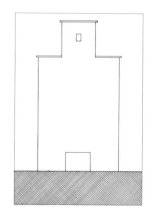

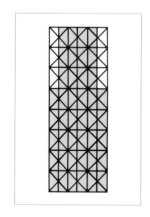
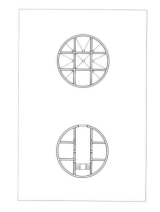
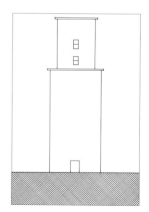

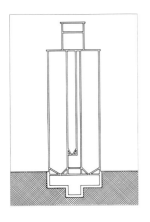

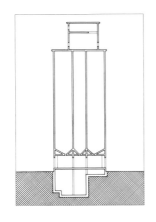

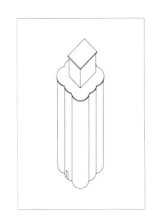

The Rural Concrete Elevator

Variation in the rural concrete elevator is found in the number of bins used. A typical elevator has an average of nine bins but the number of bins can range from one or two to more than twenty.

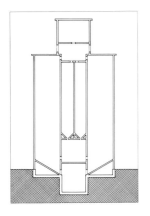

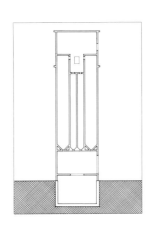

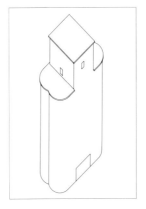

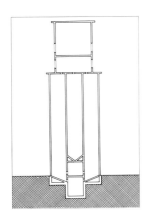

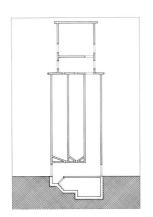

Above
Town view of Danville, Kansas where the elevators and the church are the two primary elements
Bottom row, from left to right
Medicine Lodge, Kansas; Rago, Kansas; Vesper, Kansas; Nashville, Kansas; Beeler, Kansas; Brenham, Kansas; and an abandoned square-binned elevator in York, Nebraska

TOWN PLANNING

THE RURAL ELEVATOR

Town Creation

The westward expansion of the railroad in the late 1800s was accompanied by a surge in building track-side towns, created and planned by the railroad companies in an effort to generate traffic along their lines. Towns were spaced at regular intervals determined by the amount of business the railroad predicted it could generate. This amount was calculated according to the number of productive farmland acres in the area. Companies were careful not to locate the towns too close together, for a town could fail from lack of business, or too far apart, for competing railroads might bring new lines and towns. Towns were laid out in a simple grid and planned entirely before any lots were sold. Building was restricted to the area defined by the town blocks which resulted in a sharp division between the edge of town and the surrounding farmland. Often, plans were repeated from town to town using the same block patterns and even the same street names. The grain elevators and railroad station were usually the first structures in the town to be constructed.

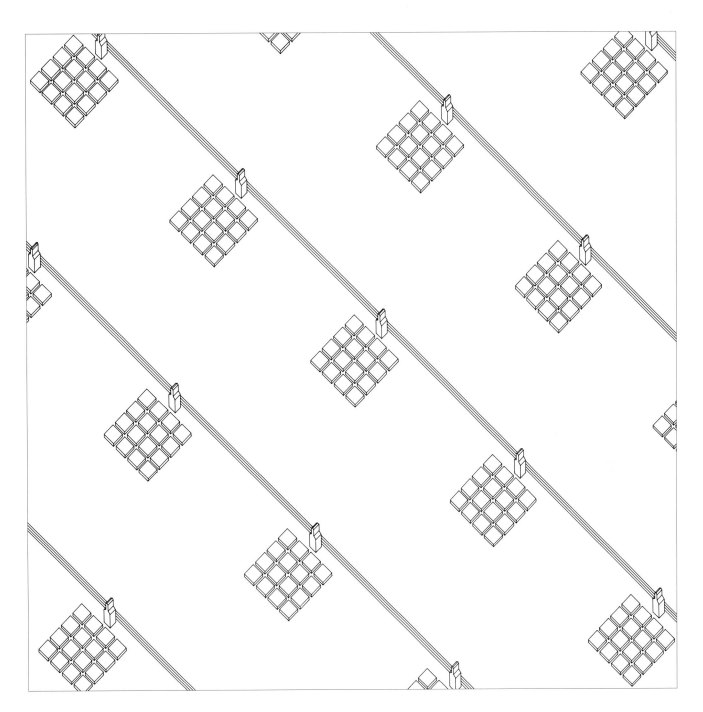

Town Orientation

Towns were oriented either in accordance with the land survey grid or along the axis of the railroad, depending primarily on whether the town formed before or after the railroad. It is common to find a combination of orientations in older towns that continued to grow. For example, the town of Galva, Illinois (opposite, top left) existed before the railroad and had streets aligned with the land grid. When the railroad subsequently came through the town on a diagonal, the center of town had to be redesigned to accommodate it.

Right
Aerial view of Sublette, Kansas
Below, from left to right
Diagrams showing various relationships between the town, the railroad, and the land grid
1 Town oriented to land grid first, railroad second
2 Town oriented to railroad first, land grid second
3 Town oriented to railroad with town additions following land grid
4 Town oriented to railroad and land grid
5 Town oriented to railroad first, land grid second
Opposite, clockwise from top left
Aerial views of Galva, Illinois; Volga, South Dakota; Benson, Minnesota; and Maxbass, North Dakota

Town Types

Almost all plains towns have a primary business street, a railroad line, an elevator, and residential blocks. Within that structure are found four primary types of towns, each identified by the different relationships between these elements and arranged in a grid pattern. These four types are known as: the symmetrical town, the orthogonal town, the T-town, and the center square town.

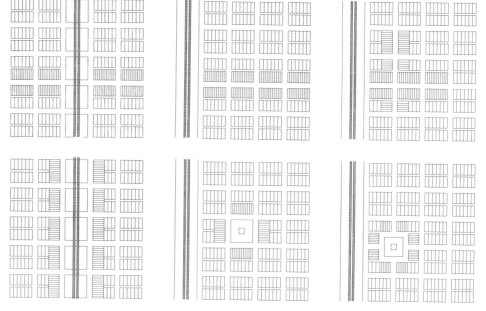

Top
Town view of Sawyer, Kansas
Center, left to right
Orthogonal town, T-town, cross T-town
Bottom, left to right
Symmetrical town, center square town, and variation of center square town
Opposite, top
Town view of Velva, North Dakota
Opposite, bottom
Town view of Surrey, North Dakota

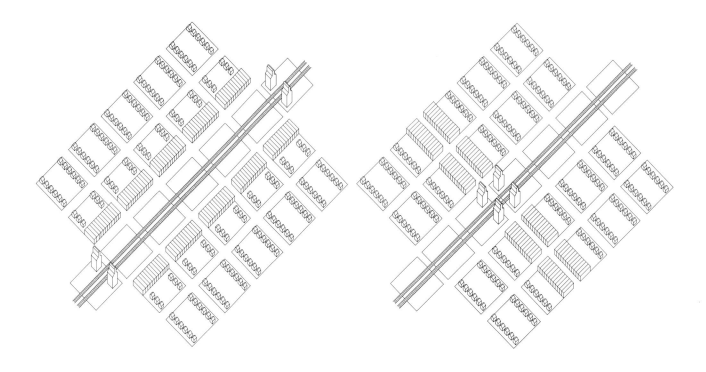

Symmetrical

The first type introduced by the railroad was the symmetrical town. Railroad tracks pass through the center of town, dividing the main street into two strips which face each other across the tracks. Railroad

companies hoped to promote the railroad and encourage its presence as a central part of the town. The symmetrical town soon lost favor with the townspeople, however, for reasons of safety. Few towns of this type remain as most were transformed into other, more workable types. Of the few remaining symmetrical towns, such as Benson, Minnesota (photograph left), most separate the tracks from the main street with narrow rows of trees that both diminish the impact of the passing trains and fulfill a community need for public park space.

Orthogonal

The orthogonal town has one main street with commercial structures on both sides of it. This main street is perpendicular to the railroad tracks and continues across them. Orthogonal towns tended to de-

velop more on one side of the tracks than on the other, which was considered a disadvantage. Although the organization of the town types varied, the size of the blocks was consistent. The standard railroad block for most towns was 300 feet square with lots 140 feet deep and backed by a twenty-foot alley lined with smaller, secondary structures. Blocks were divided into six residential lots or twelve business lots. The lots on the main street were long, narrow, and uniform, thereby creating a dense and easily identifiable spine to the town.

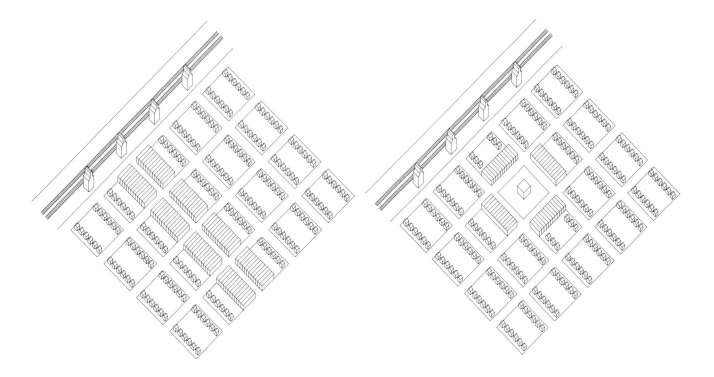

T-town

The third, and newer type is the T-town. In this type, the main street runs perpendicular to the railroad tracks but, unlike the orthogonal town, does not continue across the tracks. This plan became the preferred type for several reasons. First, it resolved the problem of the town developing on one side more than the other; and second, the tracks were moved to the edge of town where they were out of view. Further, the danger of having a railroad crossing in the center of town—one of the chief problems of the symmetrical town—was removed. A variation of the T-town, called the cross T-town, was introduced in larger towns where a primary intersection was desirable. To achieve this, a second main street was established perpendicular to the first. The crossed T-form encouraged commercial development at a central point in town, separating businesses from residences. Lots at the intersection could be sold for higher prices. Banks became the most common type of building to be placed at this intersection. Although the focus of the town gradually moved away from the railroad, the grain elevator retained its prominence simply because of its size.

Center Square Town

The center square town type developed independently from the railroad, having almost no structural relationship to it or the local elevator. This type derives from eastern town types which also evolved  separately from the construction of the railroad. Although the type provides the town with a central square, this center is often difficult to find because the main street does not extend through the town as it does in the other types. In many of the towns the center square is used as a park (an uncommon feature in most midwestern towns) or, more often, to provide a focal point for a civic building or a bank. Variations of this type yield slightly different configurations of the central square and its relationship to the existing grid of the town.

WOOD

THE URBAN ELEVATOR

The Wood Urban Elevator

Wood urban elevators were built until the end of the nineteenth century when new fire-resistant materials were introduced, and were always built with cribbed construction. Most wood elevators were covered with metal to make them less susceptible to fire, but even with this additional protection, the average life span of a wood elevator was only eleven years. The elevator on the previous page is the Shoreham elevator, located in Minneapolis, Minnesota. Built in 1894, it is one of the few urban wood elevators left.

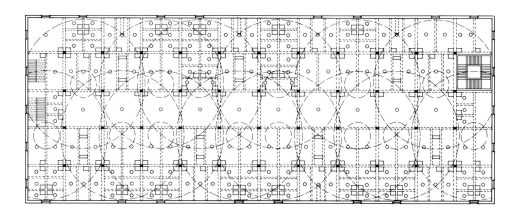

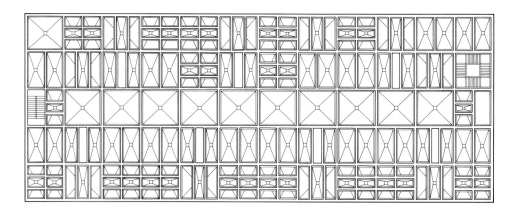

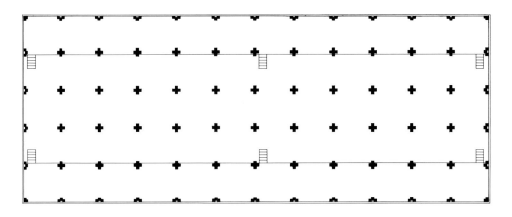

BRICK, TILE, AND STEEL

THE URBAN ELEVATOR

The Brick Urban Elevator

The brick elevator emerged in the late 1800s, at approximately the same time as the steel and tile elevators, but was the most uncommon of the three types. Brick elevators had bins that were either rectangular or circular and were generally constructed with ordinary bricks, although curved bricks were also used. All brick elevators used steel to reinforce the relatively low tensile strength of the material. The rectangular-binned elevator shown here (above right and opposite) was built in 1908 and was said to be the largest brick elevator in the world at that time; it is now being used as a hotel.

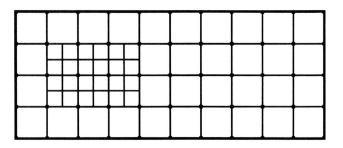

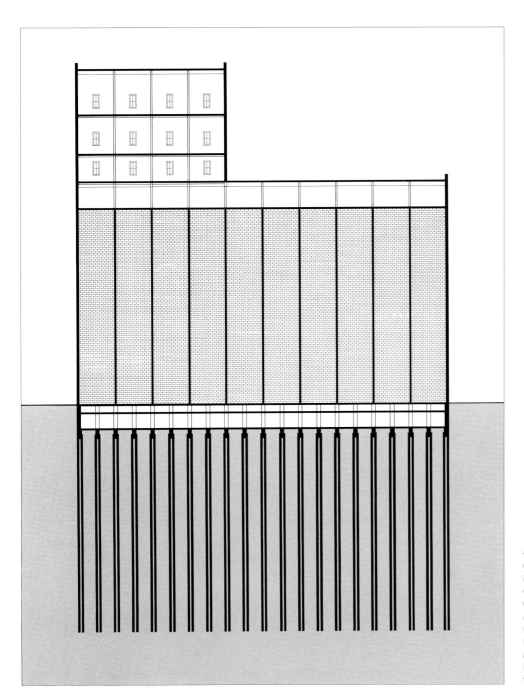

Opposite, top left
Circular-binned brick elevator
in Alton, Missouri
Opposite, top right
Ceresota elevator in Minneapolis,
Minnesota, now a hotel
Opposite, bottom
Bin plan of Ceresota elevator
Left
Longitudinal section through
the Ceresota elevator

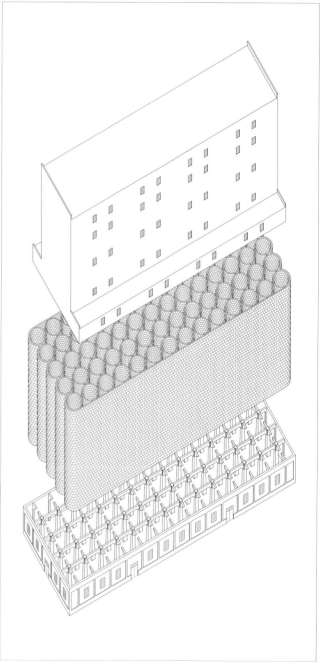

The Tile Urban Elevator

Tile was experimented with briefly between the years 1900–1915 and then, like wood, brick, and steel, was replaced by concrete. Like its rural counterpart, the urban tile elevator was built with circular bins since tiles were prefabricated with a fixed radius. Consequently, the variety of bin sizes was greatly limited, and further, many structural problems resulted from the way the bins were constructed. Tile bins were treated as individual elements rather than as a total unit. Even though most tile bins were constructed with double walls, they were almost never watertight. The elevator in the photographs above was built in 1901 in Minneapolis, Minnesota; it is one of the last urban tile elevators to survive. Its owners plan to demolish it in the near future.

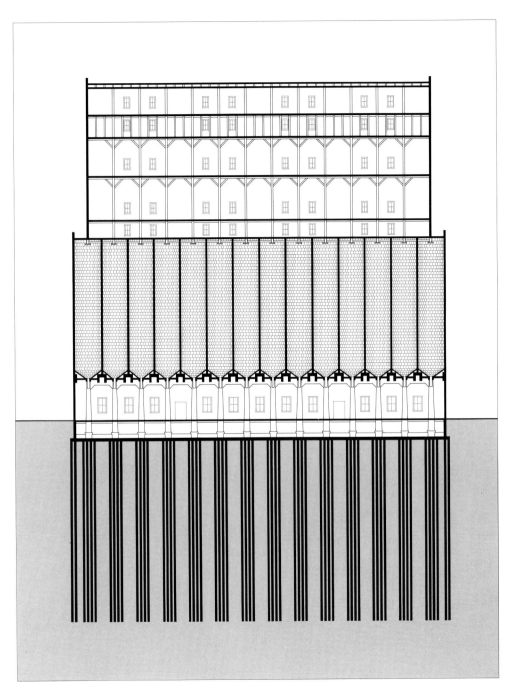

Opposite, left (top and bottom)
St. Anthony number 3 tile elevator in
Minneapolis, Minnesota
Opposite, right
Exploded axonometric of Peavey tile
elevator in Duluth, Minnesota
Left
Longitudinal section of Peavey tile
elevator

The Steel Urban Elevator

Steel elevators were constructed with either circular or rectangular bins. The building shown here has circular bins encased by a brick box to protect the bins from exposure to extreme weather conditions. Built in 1897 by the engineer Max Toltz, this elevator is the only one of its type to survive in America—its owners are currently considering demolition. Steel elevators were also built with exposed bins, and although this reduced the cost of construction, it only exacerbated the problem of insulation.

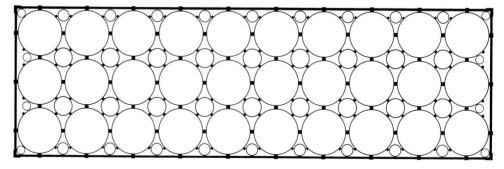

Top
View along the canal side of the Great Northern elevator showing moveable marine leg and barge
Right
Plan of bins
Opposite
Axonometric looking up at steel bin bottoms

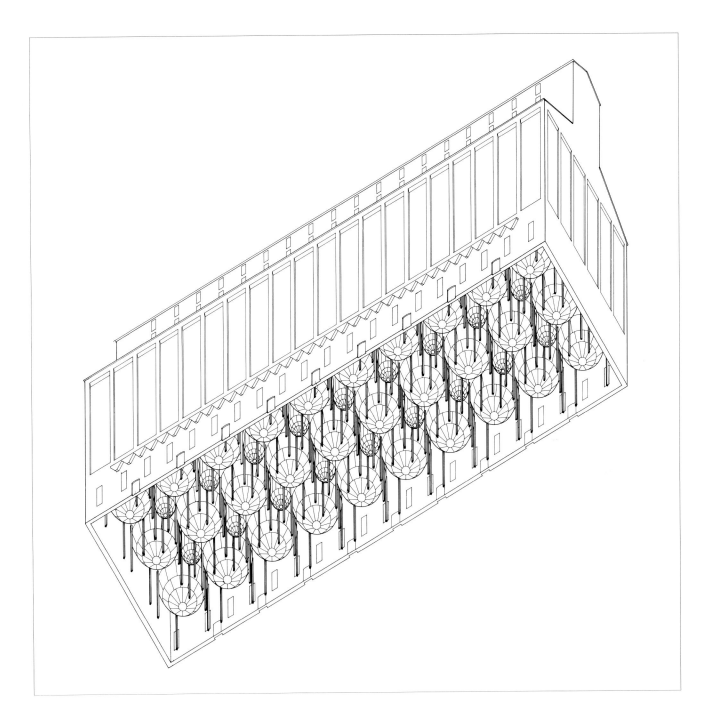

Steel Construction

Steel was a popular material for all types of construction at the end of the nineteenth century. In grain elevator construction, there were several drawbacks to using steel, however, that eventually led to its rejection in favor of concrete. First, steel was expensive and skilled labor hard to find. Second, the bins were susceptible to rust and corrosion, even when protected by an exterior box. Third, placing circular bins in a rectangular building without wasting space between them was almost impossible. Rectangular bins solved this problem, but at the expense of structural stability. Finally, steel was a poor thermal insulator, unable to provide complete protection of the grain under extreme weather conditions.

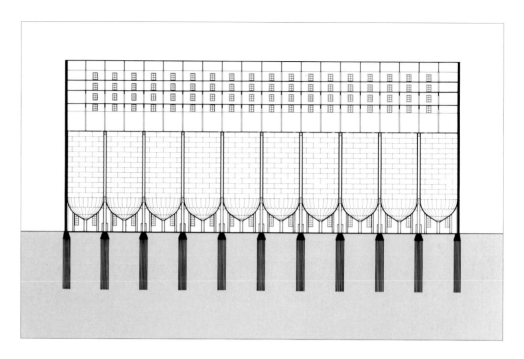

Top
Longitudinal section through the Great Northern elevator
Right
Work floor of the Great Northern elevator
Opposite
Distributing floor of the Great Northern elevator

CONCRETE

THE URBAN ELEVATOR

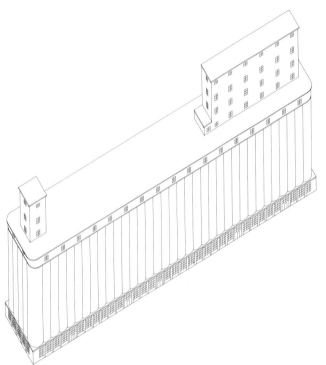

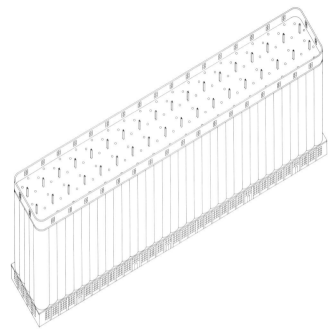

The Concrete Urban Elevator

The function of the urban elevator is to store the grain it has received from smaller rural elevators until it can be sold to processors. A typical urban elevator is made of concrete and has four distinct levels which

are easily identified from the exterior: the work floor, located at ground level; the storage bins, defining the bulk of the building; the distributing floor, directly above the bins; and the headhouse, the isolated structure on top of the building. The concrete elevator illustrated in the four drawings above is in Brooklyn, New York and was built in 1922. It is 408 feet long, 70 feet wide, and 110 feet high. It has 54 circular bins, 34 star-shaped bins, and 38 outer bins. The structure was poured as a monolith over a period of fourteen days by over 300 workers.

Distributing Floor

Located just above the main bin structure, the distributing floor is easily identified from the exterior by the large windows which span across its length. These windows provide ventilation to an area prone to

grain dust explosions. After the grain has been weighed and sorted in the headhouse it is organized and distributed into the bins here. The grain is carried along the length of the floor by conveyor belts until it reaches the appropriate bin. It is then directed down into a small hole in the floor by a rotating spout. Bin contents are recorded on a chalkboard (see opposite page, bottom left) that lists specific information about the grain: the type and amount, the company to which it belongs, the date it was stored, and the bin number.

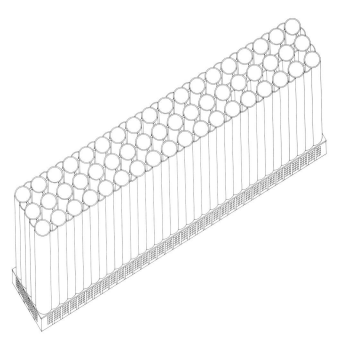

Bins

The bins comprise the storage area of the elevator. The most important features of a bin are its shape, wall configuration, and bottom slope. Concrete elevator bins are usually circular, though other forms can be found. Circular-binned elevators must support a variety of bin sizes within a single structure in order to utilize the space between cylinders economically. Such an arrangement is also advantageous for the flexibility it offers in storing varying lots of grain. The resulting odd-shaped bins within the block are called interstitial bins; those along the edges are called outerstitial bins. The bottom of a bin may be either flat or hoppered. While a flat-bottomed bin is difficult to empty thoroughly, a hoppered-bottomed bin is sloped to facilitate the movement of grain.

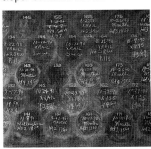

Work Floor

The work floor is located below the storage bins at ground level and connects the elevator to the train, barge, and truck loading areas. Like the distributing floor, the work floor is an open space with windows on all sides. It is distinguished, however, by the presence of the hoppered bottoms of the bins on its ceiling and the immense columns which support the structure. In the elevator illustrated above, 216 columns are spaced at 12-foot intervals. The floor—a thick concrete slab—rests on wood pylons that extend approximately 70 feet into the ground. Pylons were essential for urban elevators on or near the water where the land was marshy and unsuitable for the tremendous weight of the building. The pylons can be seen in the section on the following page.

The New Material

Concrete became the favored material for the construction of urban elevators around 1915 after twenty years of experimentation with other materials. Since then, it has been used almost exclusively. Concrete has proved to be an economical, structurally stable, fire-resistant material that provides excellent thermal protection for grain. Many early examples of the concrete urban elevator type are still standing because of the high cost of demolition—ironically, it often exceeds the initial cost of construction.

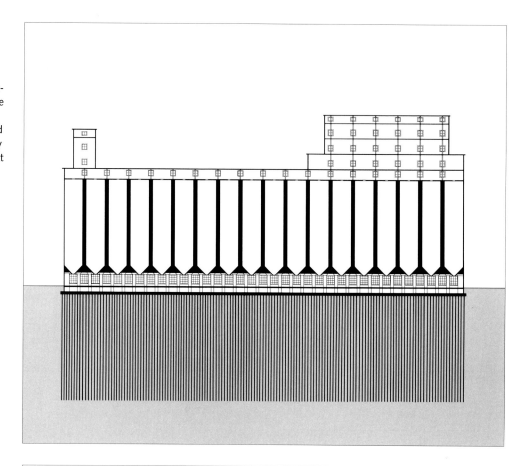

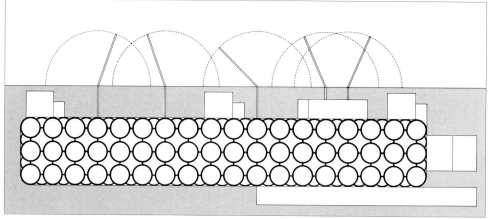

Top
Longitudinal section through Port Authority Terminal in Brooklyn, New York
Right
Site plan showing moveable marine legs and canal
Opposite
View of urban elevators in Buffalo, New York

Interiors

The simple exterior of the urban elevator is in direct contrast to the interior which more readily reflects the complex system of grain distribution and storage. The images on these two pages show the interiors of two areas of the elevator: the headhouse and the distributing floor.

Right
View inside headhouse of concrete elevator in Buffalo, New York
Opposite, top
View of distributing floor showing conveyors and holes into bins below
Opposite, center
Plan of distributing floor
Opposite, bottom
Plan of steel reinforcing at base of bins

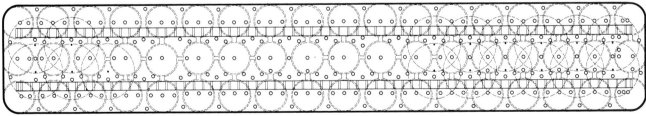

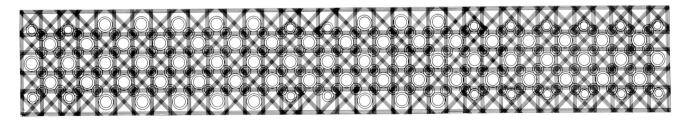

Variations in Plan

There are more bin shape variations in the concrete urban elevator than in any other type. Bins can be either circular, rectangular, hexagonal, octagonal, or a combination of these forms. Geometric variations were introduced in a attempt to find more efficient and economical storage solutions. The elevator shown in the photographs here is located in Hutchinson, Kansas and is the longest in the world. Its designers adopted the hexagonal structure of a honeycomb, allowing the space-efficient interlocking of bins of identical size (see plan and elevation on pages 14–15, figure 11).

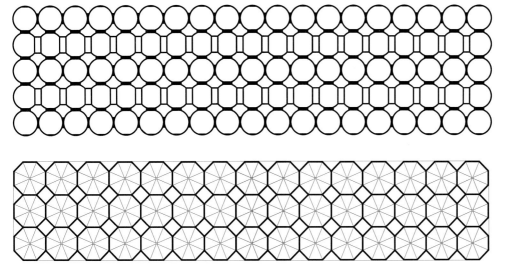

CITY PLANNING

THE URBAN ELEVATOR

City Planning

The urban elevator is not as integral to the organization of the city as the rural elevator is to the town. The urban elevator is generally placed along the water's edge—for instance, a river that runs through the city—in order to facilitate access to ships and barges.

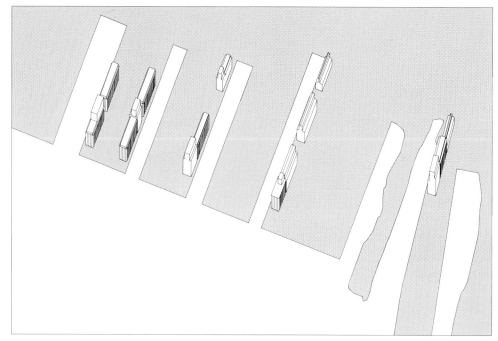

Above
View of Duluth, Minnesota
Right
Axonometric of Duluth elevators
Opposite, top
View of Buffalo, New York
Opposite, bottom
Axonometric of Buffalo elevators
(after Reyner Banham)

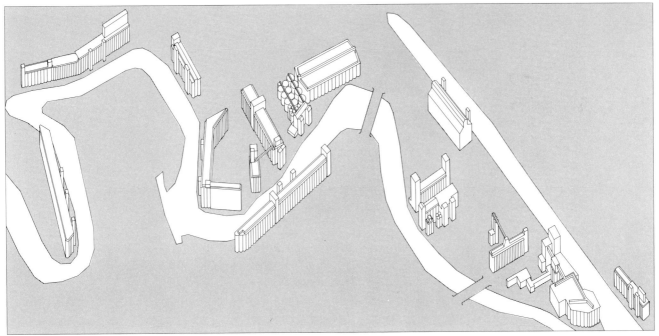

Publisher's Afterword

What is the appeal—for architects and non-architects alike—of the grain elevator? We are indebted to Lisa Mahar-Keplinger for describing this now century-old fascination, and for documenting a rich variety of elevators in both drawing and photograph. As architects, we marvel at the plans and sections seen here, kinetic in their density, yet replete with functional clarity and almost organic symmetries. As non-architects, we are moved by the elevator's presence in the photographs, from the stoic mass of the urban concrete type to the stark isolation of the rural wood elevator on the horizon.

This typological study, following the work of architects Aldo Rossi and Steven Holl, demonstrates the potential, and limitation, of this type of analysis. For Rossi, the evolution of a building type is in some way arbitrary; the function of the building is less interesting than its form, which is quite literally the vessel of meaning. Memory, tradition, and, ultimately, nostalgia are the essence of architectural typology. Holl reintroduces the idea of function into his analysis; while form may not follow function necessarily, neither can the two be divorced.

The grain elevator works both ways. On the one hand, it is one of the clearest examples of the machine aesthetic of form following function; on the other, it is one of the most potent images on the American landscape. Le Corbusier saw the concrete elevators of the Great Lakes as the triumph of American technical ingenuity and new materials; David Plowden juxtaposed the grain elevator and the cemetery as two images of decay.

The grain elevator developed following the laws of physics, seeking to avoid explosions caused by the internal pressure of grain storage; the platonic cylinder simply offered the greatest wall strength. But within this pure form resides a complex maze of conveyors, belts, chutes, pulleys, scales, and bins. The elevators are transformed, however, by construction materials (wood, brick, tile, steel, concrete), the appendage of additional storage bins, physical location, and a host of other determinants that modify the geometric purity of the Ur-form extolled by modern architects and painters.

As most of us are unfamiliar with the inner workings of the grain elevator, we respond to them largely as symbol. Although some of the large concrete elevators admired by Le Corbusier and others are still in use, most have fallen into disrepair, taking on a romantic air, not unlike the ruins of a European cathedral. Now graying, walls fissured or crumbling, these massive cathedrals of agriculture have become monuments to the golden age of American farming. Similarly, the rural grain elevator, rising above the prairie, acts as tribute (or memorial) to the small farm.

It is this dual reading—as structure and monument—that provides common ground for a wide variety of enthusiasts. The grain elevator is at once engineering wonder—a gritty machine, functionally determined; a complex embodiment of the realities of the American systems of farm production and transportation; and a symbol, be it for the passing of the family farm, the death of urban waterfronts, or a lost, naive modernism. The grain elevator is neither form, nor function, nor symbol. It is all three, arguably the most important lesson held in this enlightening book.

Kevin Lippert
New York

Bibliography

Banham, Reyner. *A Concrete Atlantis: U.S. Industrial Building and European Modern Architecture, 1900–1925*. Cambridge, MA: The MIT Press, 1986.

Clark, Schmucker, ed. *Grain Elevators of North America*. Fifth Edition. Chicago: Grain and Feed Journals Consolidated, 1942.

Clark, Schmucker, ed. *Plans of Grain Elevators*. Fourth edition. Chicago: Grain Dealers Journal, 1918.

Conzen, Michael, ed. *The Making of the American Landscape*. Boston: Unwin Hyman, 1990.

Frame, Robert M. III. "Grain Elevators in Minnesota to 1945." Multiple Property Documentation Form, National Register of Historic Places, 1989.

Holl, Steven. *Rural and Urban House Types in North America*. New York: Pamphlet Architecture, 1982.

Hudson, John. *Plains Country Towns*. Minneapolis: University of Minnesota Press, 1985.

Marx, Leo. *The Machine in the Garden: Technology and the Pastoral Ideal in America*. New York: Oxford University Press, 1964.

Morris, Wright. *Time Pieces: Photographs, Writing, and Memory*. New York: Aperture, 1989.

Reps, John W. *The Making of Urban America: A History of City Planning in the United States*. Princeton: Princeton University Press, 1965, pp. 382–400.

Stott, William. *Documentary Expression and Thirties America*. Chicago: The University of Chicago Press, 1973.

Tsujimoto, Karen. *Images of America: Precisionist Painting and Modern Photography*. Seattle: University of Washington Press, 1982.

Photography Credits

Jet Lowe, Historic American Engineering Record
All photographs taken in Buffalo, New York, December 1990
5 (top, fourth down, fifth down), 59, 64, 66, 67, 70 (bottom left and right), 71 (bottom left and right), 73, 74, 75, 81

Arthur Rothstein, Farm Security Administration
46 (Fairfield, Montana, 1939)

John Vachon, Farm Security Administration
13 and 80 (Duluth, Minnesota, 1941),
51 (top: Velva, North Dakota, 1940),
51 (bottom: Surrey, North Dakota, 1940),
79 (St. Paul, Minnesota, 1939)

Marion Post Wolcott
Farm Security Administration
12 (Carter, Montana, 1941)

All other photographs taken by the author between April 1991 and May 1992

Acknowledgments

First, I would like to thank all those who gave me the opportunity to make this catalogue. The National Endowment for the Arts and the New York State Council on the Arts provided me with the funding for this study, giving me opportunities I never would have had without its assistance. Thanks also to the Preservation Coalition of Erie County who supported my initial idea for the project and helped me to acquire the funding to carry it through. Kevin Lippert of Princeton Architectural Press believed in the project from the beginning and agreed to publish the study even though what he first saw was rough and vague. I am grateful that he was able to take the time to write the afterword which is written with intelligence and care.

With great respect and admiration, I must also thank Aldo Rossi, who in his moving introduction, captured the poetry of these buildings and their relationship to the American landscape with an unequalled sensitivity that is uniquely his own. He has given me many opportunities in his office, inspiring me with his passion for this type of architecture and his ability to see spirituality in simple forms. A special thanks to his partner Morris Adjmi, who supported and encouraged me through all the ups and downs, urging me to pursue the study even though it frequently conflicted with my responsibilities at the office.

I am particularly indebted to Virginia Polytechnic Institute and State University for the education I received there; and especially to Professors Olivio Ferrari, Ron Daniel, and Gene Egger for their faith and continued support.

Finding the information necessary to begin this study was difficult because of the lack of previous documentation on the subject. Robert Frame III was invaluable to the study. He pointed me in the right direction and generously shared his own research. His suggestion to read John Hudson's book *Plains Country Towns* generated the inclusion of the section on town development. I am indebted to both scholars for their important observations and research. Eric DeLony of the Historic American Engineering Record further encouraged me to pursue this study and introduced me to the beautiful photographs by Jet Lowe, a number of which are included in this catalogue. Through HAER I also met Craig Strong, who kept me informed of the organization's research of the Buffalo grain elevators they were documenting in 1990.

Second, I must thank those who helped me produce this catalogue. Jim Biek drafted the majority of the drawings with great skill and sensitivity. I am very grateful for his involvement. Sang Lee and Adam Cohen helped me with drawings in the beginning as did Ted Greenleaf, who also helped in developing the initial organization of the study. Stefanie Lew of Princeton Architectural Press spent a good deal of time editing the text and finalizing the pages. She helped me make many difficult decisions and was meticulous in preparing the final manuscript. Many thanks to Diane Ghirardo for her beautiful translation of Aldo Rossi's "Timeless Cathedrals," and to Steven Holl and my good friend Karen Stein who took the time to look at the catalogue carefully before its publication and made many valuable criticisms. I must also thank my close friends Nancy Herrmann, for accompanying me on several photography trips; and Ana Marton, for her vision and compassion. Most of all, I must thank my husband, Eric Keplinger; without his support this book would not have been possible.